Summary

The earthquake and subsequent tsunami that devastated Japan's Fukushima Daiichi nuclear power station and the earthquake that forced the North Anna, VA, nuclear power plant's temporary shutdown have focused attention on the seismic criteria applied to siting and designing commercial nuclear power plants. Some Members of Congress have questioned whether U.S nuclear plants are more vulnerable to seismic threats than previously assessed, particularly given the Nuclear Regulatory Commission's (NRC's) ongoing reassessment of seismic risks at certain plant sites.

The design and operation of commercial nuclear power plants operating in the United States vary considerably because most were custom-designed and custom-built. Boiling water reactors (BWRs) directly generate steam inside the reactor vessel. Pressurized water reactors (PWRs) use heat exchangers to convert the heat generated by the reactor core into steam outside of the reactor vessel. U.S. utilities currently operate 104 nuclear power reactors at 65 sites in 31 states; 69 are PWR designs and the 35 are BWR designs.

One of the most severe operating conditions a reactor may face is a loss of coolant accident (LOCA), which can lead to a reactor core meltdown. The emergency core cooling system (ECCS) provides core cooling to minimize fuel damage by injecting large amounts of cool water containing boron (borated water slows the fission process) into the reactor coolant system following a pipe rupture or other water loss. The ECCS must be sized to provide adequate make-up water to compensate for a break of the largest diameter pipe in the primary system (i.e., the so-called "double-ended guillotine break" (DEGB)). The NRC considers the DEGB to be an extremely unlikely event; however, even unlikely events can occur, as the magnitude 9.0 earthquake and resulting tsunami that struck Fukushima Daiichi proves.

U.S. nuclear power plants designed in the 1960s and 1970s used a deterministic statistical approach to addressing the risk of damage from shaking caused by a large earthquake (termed Deterministic Seismic Hazard Analysis, or DSHA). Since then, engineers have adopted a more comprehensive approach to design known as Probabilistic Seismic Hazard Analysis (PSHA). PSHA estimates the likelihood that various levels of ground motion will be exceeded at a given location in a given future time period. New nuclear plant designs will apply PSHA.

In 2008, the U.S Geological Survey (USGS) updated the National Seismic Hazard Maps (NSHM) that were last revised in 2002. USGS notes that the 2008 hazard maps differ significantly from the 2002 maps in many parts of the United States, and generally show 10%-15% reductions in spectral and peak ground acceleration across much of the Central and Eastern United States (CEUS), and about 10% reductions for spectral and peak horizontal ground acceleration in the Western United States (WUS). Spectral acceleration refers to ground motion over a range, or spectra, of frequencies. Seismic hazards are greatest in the WUS, particularly in California, Oregon, and Washington, as well as Alaska and Hawaii.

In 2010, the NRC examined the implications of the updated NSHM for nuclear power plants operating in the CEUS, and concluded that NSHM data suggest that the probability for earthquake ground motions may be above the seismic design basis for some nuclear plants in the CEUS. In late March 2011, NRC announced that it had identified 27 nuclear reactors operating in the CEUS that would receive priority earthquake safety reviews.

Contents

Figures

Tables

Appendixes

Contacts

Background

The seismic design criteria applied to siting commercial nuclear power plants operating in the United States received increased attention following the March 11, 2011, earthquake and tsunami that devastated Japan's Fukushima Daiichi nuclear power station. Since the event, a magnitude 5.8 earthquake near Mineral, VA, on August 23, 2011, precipitated the temporary shutdown of Dominion Power's North Anna nuclear power plant. Some Members of Congress have questioned whether U.S nuclear plants are more vulnerable to seismic threats than previously assessed, particularly given the Nuclear Regulatory Commission's (NRC's) ongoing reassessment of seismic risks at certain plant sites.[1]

Currently, 104 commercial nuclear power plants operating in the United States use variations in light water reactor designs and construction. **Figure 1** shows the locations of all 104 nuclear power reactors operating in the United States.

Light water reactors use ordinary water as a neutron moderator and coolant, and uranium fuel enriched in fissile uranium-235.[2] Designs fall into either pressurized water reactor (PWR) or boiling water reactor (BWR) categories. Both have reactor cores (the source of heat) consisting of arrays of uranium fuel bundles capable of sustaining a controlled nuclear chain reaction.[3] U.S. commercial nuclear power plants incorporate safety features intended to ensure that, in the event of an earthquake, the reactor core would remain cooled, the reactor containment would remain intact, and radioactive releases would not occur from spent fuel storage pools. NRC defines this as the "safe-shutdown condition."

When utilities began building nuclear power plants in the 1960s-1970s era, they typically hired an architect/engineering firm, then contracted with a reactor manufacturer ("nuclear vendors") to build the nuclear steam supply system (NSSS), consisting of the nuclear core, reactor vessel, steam generators and pressurizer (in PWRs), and control mechanisms—representing about 10% of the plant investment.[4] The balance of the plant (BOP) consisted of secondary cooling systems, feed-water systems, steam systems, control room, and generator systems. At the time, the four vendors who offered designs for nuclear reactor systems in the United States were Babcock & Wilcox, Combustion Engineering, General Electric, and Westinghouse. About 12 architect/engineering firms were available to design the balance of the plant. Each architect/engineer had its own preferred approach to designing the balance of plant systems. The custom design-and-build industry approach resulted in problems verifying the safety of individual plants and in transferring the safety lessons learned from one reactor to another. In addition to the custom-design features of each plant, designers also had to contend with earthquake hazards unique to each plant site. Designs for structures, systems, and components important to a nuclear power plant operation must withstand earthquakes without losing their intended safety-related function.

---

[1] This report does not discuss the risk from earthquake-caused tsunamis, as associated with the catastrophic damage to the Fukushima plants.

[2] Heavy water reactors, such as Canada's CANDU reactor, use water containing a heavier hydrogen isotope and natural uranium for fuel, which contains about 0.7% uranium-235.

[3] For further background uranium fuel, see CRS Report RL34234, *Managing the Nuclear Fuel Cycle: Policy Implications of Expanding Global Access to Nuclear Power*, coordinated by Mary Beth Nikitin.

[4] Office of Technology Assessment, *Nuclear Power Plant Standardization: Light Water Reactors*, NTIS order #PB81-213589, April 1981, p. 11.

This report presents some of the general design concepts of operating nuclear power plants in order to discuss design considerations for seismic events. This report does not attempt to conclude whether one design is inherently safer or less safe than another plant. Nor does it attempt to conclude whether operating nuclear power plants are at any greater or lesser risk from earthquakes given recent updates to seismic data and seismic hazard maps.

Figure 1. Commercial Nuclear Power Plants Operating in the United States

(One hundred and four [104] Operating Reactors)

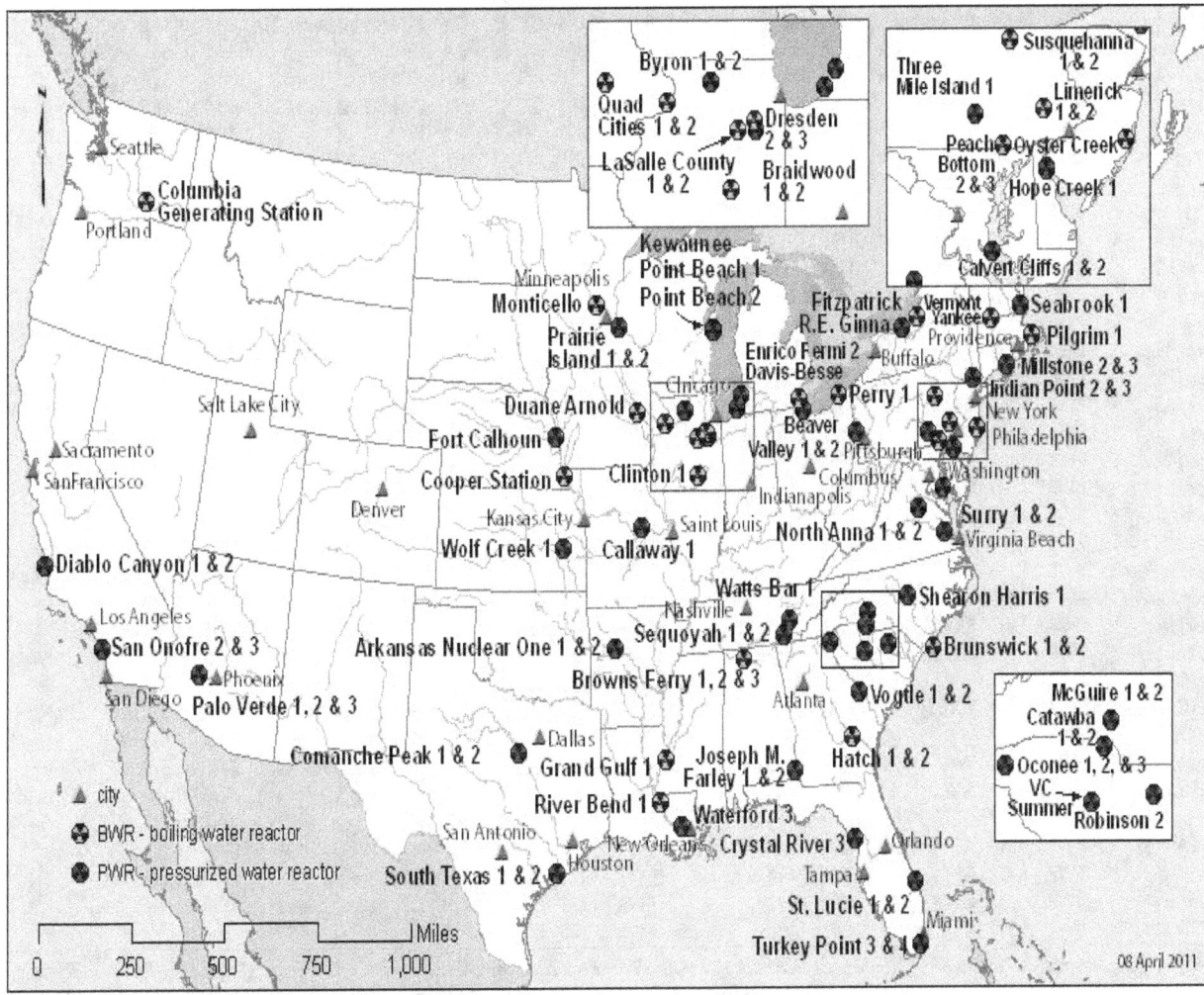

Source: Prepared by the Library of Congress Geography and Maps Division for CRS using U.S. NRC Find Operating Nuclear Reactors by Location or Name, http://www.nrc.gov/info-finder/reactor/index.html#AlphabeticalList.

Notes: Currently, 104 nuclear power reactors operate at 65 sites in 31 states; 69 are PWR designs and the 35 remaining are BWR designs.

Notes:

Unit	Type	MW	Vendor	St.	Lic.	Unit	Type	MW	Vendor	St.	Lic.	Unit	Type	MW	Vendor	St.	Lic.
Arkansas Nuclear 1	PWR	843	B&W	AK	1974	Grand Gulf 1	BWR	1,297	GET6	MS	1984	Point Beach 1	PWR	512	W2L	WI	1970
Arkansas Nuclear 2	PWR	995	CE	AK	1974	Hatch 1	BWR	876	GET4	GA	1974	Point Beach 2	PWR	514	W2L	WI	1973
Beaver Valley 1	PWR	892	W3L	PA	1976	Hatch 2	BWR	883	GET4	GA	1978	Prairie Island 1	PWR	551	W2L	MN	1874
Beaver Valley 2	PWR	846	W3L	PA	1987	Robinson 2	PWR	710	W3L	SC	1970	Prairie Island 2	PWR	545	W2L	MN	1974
Braidwood 1	PWR	1,178	W4L	IL	1987	Hope Creek 1	BWR	1,061	GET4	NJ	1986	Quad Cities 1	BWR	867	GET3	IL	1972
Braidwood 2	PWR	1,152	W4L	IL	1988	Indian Point 2	PWR	1,023	W4L	NY	1973	Quad Cities 2	BWR	869	GET3	IL	1972
Browns Ferry 1	BWR	1,065	GET4	AL	1973	Indian Point 3	PWR	1,025	W4L	NY	1975	R. E. Ginna	PWR	498	W2L	NY	1969
Browns Ferry 2	BWR	1,104	GET4	AL	1974	Joseph M. Farley 1	PWR	851	W3L	AL	1977	River Bend 1	BWR	989	GET6	LA	1985
Browns Ferry 3	BWR	1,115	GET4	AL	1976	Joseph M. Farley 2	PWR	860	W3L	AL	1981	Salem 1	PWR	1,174	W4L	NJ	1976
Brunswick 1	BWR	938	GET4	NC	1976	Kewaunee	PWR	556	W2L	WI	1973	Salem 2	PWR	1,130	W4l	NJ	1981
Brunswick 2	BWR	937	GET4	NC	1974	LaSalle County 1	BWR	1,118	GET5	IL	1982	San Onofre 2	PWR	1,070	CE	CA	1982
Byron 1	PWR	1,164	W4L	IL	1985	LaSalle County 2	BWR	1,120	GET5	IL	1983	San Onofre 3	PWR	1,080	CE	CA	1992
Byron 2	PWR	1,136	W4L	IL	1987	Limerick 1	BWR	1,134	GET4	PA	1985	Seabrook 1	PWR	1,295	W4L	NH	1990
Callaway 1	PWR	1,236	WFL	MO	1984	Limerick 2	BWR	1,134	GET4	PA	1989	Sequoyah 1	PWR	1,148	W4L	TN	1980
Calvert Cliffs 1	PWR	873	CE	MD	1974	McGuire 1	PWR	1,100	W4L	NC	1981	Sequoyah 2	PWR	1,126	W4L	TN	1981
Calvert Cliffs 2	PWR	862	CE	MD	1976	McGuire 2	PWR	1,100	W4L	NC	1983	Shearon Harris 1	PWR	900	W3L	NC	1986
Catawba 1	PWR	1,129	W4L	SC	1985	Millstone 2	PWR	884	CE	CT	1975	South Texas 1	PWR	1,410	W4L	TX	1988
Catawba 2	PWR	1,129	W4L	SC	1986	Millstone 3	PWR	1,227	W4L	CT	1986	South Texas 2	PWR	1,410	W4L	TX	1989
Clinton 1	BWR	1,065	GET6	IL	1987	Monticello	BWR	579	GET3	MN	1970	St. Lucie 1	PWR	839	CE	FL	1976
Columbia Gen. St.	BWR	1,190	GET5	WA	1984	Nine Mile Pt .1	BWR	621	GET2	NY	1974	St. Lucie 2	PWR	839	CE	FL	1983
Comanche Peak 1	PWR	1,200	W4L	TX	1990	Nine Mile Pt. 2	BWR	1,140	GET5	NY	1987	Surry 1	PWR	799	W3L	VA	1972
Comanche Peak 2	PWR	1,150	W4L	TX	1993	North Anna 1	PWR	981	W3L	VA	1978	Surry 2	PWR	799	W3l	VA	1973
Cooper Station	BWR	830	GET4	NE	1974	North Anna 2	PWR	973	W3L	VA	1980	Susquehanna 1	BWR	1,149	GET4	PA	1982
Crystal River 3	PWR	838	B&WLL	FL	1976	Oconee 1	PWR	846	B&WLL	SC	1973	Susquehanna 2	BWR	1,140	GET4	PA	1984
Davis-Besse	PWR	893	B&WLL	OH	1977	Oconee 2	PWR	846	B&WLL	SC	1973	Three Mile Isl. 1	PWR	786	B&WLL	PA	1974
Diablo Canyon 1	PWR	1,151	W4L	CA	1984	Oconee 3	PWR	846	B&WLL	SC	1974	Turkey Point 3	PWR	720	W3L	FL	1972
Diablo Canyon 2	PWR	1149	W4L	CA	1985	Oyster Creek	BWR	619	GET2	NJ	1991	Turkey Point 4	PWR	720	W3l	FL	1973
Donald C. Cook 1	PWR	1,009	W4L	MI	1974	Palisades	PWR	778	CE	MI	1971	VC Summer	PWR	966	W3l	SC	1982
Donald C. Cook 2	PWR	1,060	W4L	MI	1977	Palo Verde 1	PWR	1,335	CES80	AZ	1985	Vermont Yankee	BWR	510	GET4	VT	1972
Dresden 2	BWR	867	GET3	IL	1991	Palo Verde 2	PWR	1,335	CES80	AZ	1986	Vogtle 1	PWR	1,109	W4L	GA	1987
Dresden 3	BWR	867	GET3	IL	1971	Palo Verde 3	PWR	1,335	CES80	AZ	1987	Vogtle 2	PWR	1,127	W4L	GA	1989
Duane Arnold	BWR	640	GET4	IA	1974	Peach Bottom 2	BWR	1,112	GET4	PA	1973	Waterford 3	PWR	1,250	CE	LA	1985
Fermi 2	BWR	1,122	GET4	MI	1985	Peach Bottom 3	BWR	1,112	GET4	PA	1974	Watts Bar 1	PWR	1,123	W4l	TN	1996
Fitzpatrick	BWR	852	GET4	NY	1974	Perry 1	BWR	1,261	GET6	OH	1986	Wolf Creek 1	PWR	1,166	W4L	KS	1985
Fort Calhoun	PWR	500	CE	NE	1973	Pilgrim 1	BWR	685	GET3	MA	1972						

Notes: No commercial nuclear power plants operate in Alaska or Hawaii. B&W: Babcock & Wilcox 2-Loop Lower; CE: Combustion Engineering; CE80: Combustion Engineering System 80; W2L Westinghouse 2-Loop; W3L Westinghouse 3-Loop; W4L Westinghouse 4-Loop; GET2: General Electric Type 2; GET3: General Electric Type 3; GET4: General Electric Type 4; GET5: General Electric Type 5; GET6: General Electric Type 6.

Nuclear Power Plant Designs

General design criteria for nuclear power plants require that structures and components important to safety withstand the effects of earthquakes, tornados, hurricanes, floods, tsunamis, and seiche waves[5] without losing the capability to perform their safety function. These "safety-related" structures, systems, and components are those necessary to assure:

- The capability to maintain the reactor coolant pressure,

- The capability to shut down the reactor and maintain it in a safe condition, or

- The capability to prevent or mitigate the consequences of accidents, which could result in potential offsite radiation exposures.

All BWR plants operating in the United States use variations of a General Electric design. The more numerous PWR plants use Babcock & Wilcox, Combustion Engineering, and Westinghouse designs. **Table 1** summarizes the various reactor types. The sections that follow discuss them further.

Table 1. Reactor Type, Vendor, and Containment

Type	Vendor	Containment	No. of operating reactors.
BWR	General Electric Type 2	Wet, Mark I	2
	General Electric Type 3	Wet, Mark I	6
	General Electric Type 4	Wet, Mark I	15
	General Electric Type 4	Wet, Mark II	4
	General Electric Type 5	Wet, Mark II	4
	General Electric Type 6	Wet, Mark III	<u>4</u>
			35
PWR	Babcock & Wilcox 2-Loop Lower	Dry, Ambient Pressure	7
	Combustion Engineering	Dry, Ambient Pressure	11
	Combustion Engineering System 80	Large Dry, Ambient Pressure	3
	Westinghouse 2-Loop	Dry, Ambient Pressure	6
	Westinghouse 3-Loop	Dry, Ambient Pressure	7
	Westinghouse 3-Loop	Dry, Sub-atmospheric	6
	Westinghouse 4-Loop	Dry, Ambient Pressure	18
	Westinghouse 4-Loop	Dry, Sub-atmospheric	1
	Westinghouse 4-Loop	Wet, Ice Condenser	9
	Westinghouse 4-Loop	Dry, Ambient Pressure	<u>1</u>
			69

Source: U.S. NRC.

Boiling Water Reactor (BWR) Systems

A boiling water reactor generates steam directly inside the reactor vessel as water flows upward through the reactor's core (see **Figure 2**).[6] The water also cools the reactor core, and the reactor

---

[5] Standing waves, or waves that move vertically but not horizontally. Seiche waves can be triggered by earthquakes, strong winds, tides, and other causes.

[6] U.S. Nuclear Regulatory Commission, *Reactor Concepts Manual*, *Boiling Water Reactor Systems*, (continued...)

---

operator is able to vary the reactor's power by controlling the rate of water flow through the core with recirculation pumps and jet pumps. The generated steam flows out the top of the reactor vessel through pipelines to a combined high-pressure/low-pressure turbine-generator. After the exhausted steam leaves the low-pressure turbine, it runs through a condenser/heat exchanger that cools the steam and condenses it back to water. A series of pumps return the condensed water back to the reactor vessel. The heat exchanger cycles cooling water through a cooling tower, or takes in water and directly discharges it to a lake, river, or ocean. The water that flows through the reactor, steam turbines, and condenser is a closed loop that never contacts the outside environment under normal operating conditions. Reactors of this design operate at temperatures of approximately 570° F and pressures of 1,000 pounds per square inch (psi) atmospheric.

Figure 2. Boiling Water Reactor (BWR) Plant

(Generic Design Features)

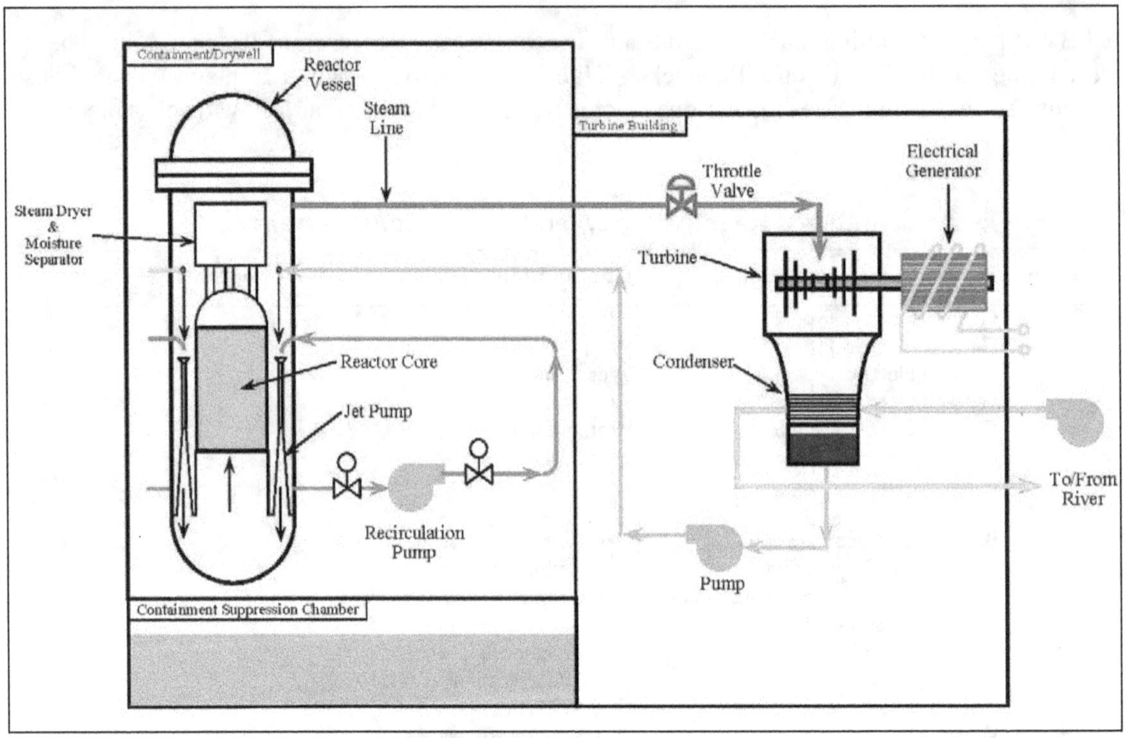

Source: U.S. Nuclear Regulatory Commission, *Reactor Concepts Manual, Boiling Water Reactor Systems,* 2005.

BWR Safe Shutdown Condition

In the case of events that cause a nuclear power plant to exceed its operating parameters (for example, an earthquake or a critical component's failure) design safety features must provide a means to control reactivity and cool the reactor.

During normal operation, reactor cooling relies on the water that enters the reactor vessel and the generated steam that exits. During safe shutdown, after the fission process is halted, the reactor

(...continued)

http://www.nrc.gov/reading-rm/basic-ref/teachers/03.pdf, October 17, 2005.

core continues to generate heat by radioactive decay and generates steam.[7] The heat from this radioactive decay initially equals about 6% of the heat produced by the reactor at full power and gradually declines. Under this condition, the steam bypasses the turbine and diverts directly to the condenser to cool the reactor. When the reactor vessel pressure decreases to approximately 50 psi, the shutdown-cooling mode removes residual heat by pumping water from the reactor recirculation loop through a heat exchanger and back to the reactor via the recirculation loop. The recirculation loop design limits the number of pipes that penetrate the reactor vessel.

Loss of Coolant Accident

The most severe operating condition affecting a BWR is a loss of coolant accident (LOCA). In the absence of coolant, the uncovered reactor core continues to generate heat through radioactive decay. The resulting heat buildup can damage the fuel or fuel cladding and lead to a fuel "meltdown." Under such a condition, an emergency core cooling system (ECCS) provides water to cool the reactor core. The ECCS is an independent high-pressure coolant injection system that requires no auxiliary electrical power, plant air systems, or external cooling water systems to provide makeup water under small and intermediate loss of coolant accidents. A low-pressure ECCS sprays water from the suppression pool into the reactor vessel and on top of the fuel assemblies.[8] The ECCS must also be sized to provide adequate makeup water to compensate for a break of the largest diameter pipe in the primary system (i.e., the so-called "double-ended guillotine break" (DEGB)). The NRC views the DEGB as an extremely unlikely event (likely to occur only once per 100,000 years of reactor operation).[9]

BWR Design Evolution

Only General Electric boiling water reactors operate in the United States (**Table 1**). BWRs are inherently simpler designs than other light water reactor types. Since they heat water and generate steam directly inside the reactor vessel, they have fewer components than pressurized water reactors. The original BWR design-types have been decommissioned, but Type 2 through Type 6 BWRs continue to operate. Some of the BWR evolutionary design features are summarized in **Table 2**. Along with the evolution in BWR reactor design, containment structure designs have also evolved (**Figure 3**, **Figure 4**, and **Figure 5**).

---

[7] During the sustained chain reaction in an operating reactor, the U-235 splits into highly radioactive fission products, while the U-238 is partially converted to plutonium-239 by neutron capture, some of which also fissions. Further neutron capture creates other radioactive elements. The process of radioactive decay transforms an atom to a more stable element through the release of radiation—alpha particles (two protons and two neutrons), charged beta particles (positive or negative electrons), or gamma rays (electromagnetic radiation).

[8] The NRC regulates the design, construction, and operation requirements of the ECCS under 10 CFR50.46, "Acceptance criteria for emergency core cooling systems for light-water nuclear reactors"; Appendix K to 10 CFR 50, "ECCS Evaluation Models"; and Appendix A to 10 CFR 50, "General Design Criteria [GDC] for Nuclear Power Plants" (e.g., GDC 35, "Emergency Core Cooling").

[9] N.C. Chokshi, S.K. Shaukat, and A.L. Hiser, et al., *Seismic Considerations for the Transition Break Size*, U.S. Nuclear Regulatory Commission, NUREG 1903, Brookhaven National Laboratory, February 2008.

Table 2. BWR Design Evolution

Model	Year Introduced	Design Feature	Typical Plants
BWR/1	1955	Natural circulation	Dresden 1
		First internal steam separation	Big Rock Point
		Isolation condenser	Humboldt Bay
		Pressure Suppression Containment	
BWR/2	1963	Large direct cycle	Oyster Creek
BWR/3/4	1965/1966	First jet pump application	Dresden 2
		Improved Emergency Core Cooling System (ECCS); spray and flood	Browns Ferry
		Reactor Core Isolation Cooling, (RCIC) system	
BWR/5	1969	Improved ECCS systems	LaSalle
		Valve recirculation flow control	9 Mile Point 2
BWR/6	1972	Improved jet pumps and steam separators	Clinton
		Reduced fuel duty: 13.4 kW/ft, 44 kW/m	Grand Gulf
		Improved ECCS performance	Perry
		Gravity containment flooder	
		Solid-state nuclear system protection system (Option, Clinton only)	
		Compact control room option	

Source: M. Ragheb, Chapter 3, *Boiling Water Reactors*, https://netfiles.uiuc.edu/mragheb/www/ NPRE%20402%20ME%20405%20Nuclear%20Power%20Engineering/Boiling%20Water%20Reactors.pdf.

Note: All BWR/1 plants that operated in the United States have been decommissioned.

Figure 3. GE BWR / Mark I Containment Structure

(Showing Torus Suppression Pool)

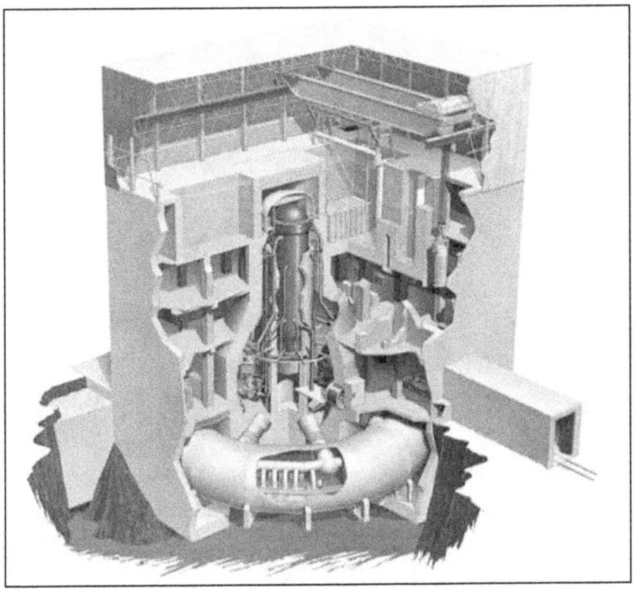

Source: General Electric, in *NRC Boiling Water Reactor (BWR) Systems*, http://www.nrc.gov/reading-rm/basic-ref/teachers/03.pdf.

Note: Japan's Fukushima Daiichi plants use designs similar to this.

Figure 4. General Electric Mark II Containment Structure

Source: General Electric, in NRC *Boiling Water Reactor (BWR) Systems*, http://www.nrc.gov/reading-rm/basic-ref/teachers/03.pdf.

Figure 5. General Electric Mark III Containment Structure

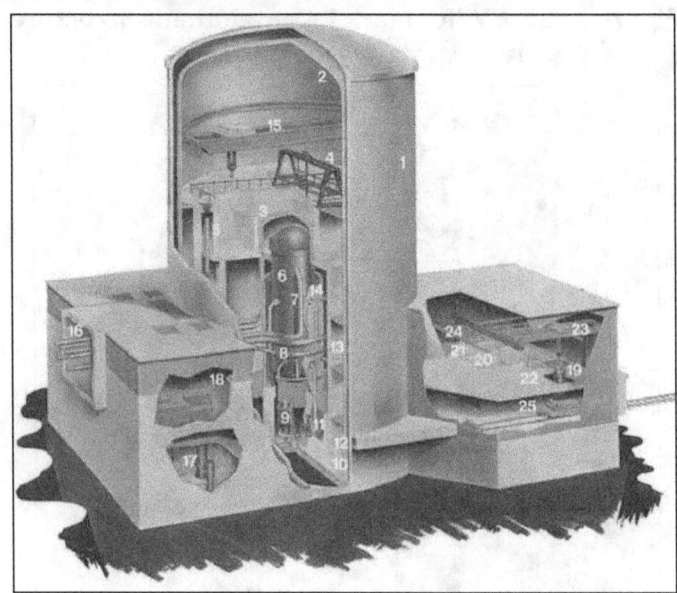

Source: General Electric, in *NRC Boiling Water Reactor (BWR) Systems*, http://www.nrc.gov/reading-rm/basic-ref/teachers/03.pdf.

Notes:

Reactor Building	Auxiliary Building	Fuel Building
1. Shield Building	16. Steam Line Channel	19. Spent Fuel Shipping cask
2. Free Standing Steel Containment	17. RHR System	20. Fuel Storage Pool
3. Upper Pool	18. Electrical Equipment Room	21. Fuel Transfer Pool
4. Refue ing Platform		22. Cask Loading Pool
5. Reactor Water Cleanup		23. Cask Handling Crane
6. Reactor Vessel		24. Fuel Transfer Bridge
7. Steam Line		25. Fuel Cask Skid on Railroad Car
8. Feed-water Line		
9. Recirculation Loop		
10. Suppression Pool		
11. Weir Wall		
12. Horizontal Vent		
13. Dry Well		
14. Shield Wall		
15. Polar Crane		

Pressurized Water Reactor Systems

A pressurized water reactor (PWR) generates steam outside the reactor vessel, unlike a BWR design. A primary system (reactor cooling system) cycles superheated water from the core to a heat exchanger/steam generator. A secondary system then transfers steam to a combined high-pressure/low-pressure turbine generator (**Figure 6**).[10] Steam exhausted from the low-pressure turbine runs through a condenser that cools and condenses it back to water. Pumps return the cooled water back to the steam generator for reuse. The condenser cools the steam leaving the turbine-generator through a third system by flowing past a heat-exchanger that recycles cooling water through a cooling tower, or takes in water and directly discharges it to a lake, river, or ocean. Unlike a BWR design, the cooling water that flows through the reactor core never contacts the turbine-generator. Under normal operating conditions, reactor cooling-water does not contact the environment.

Figure 6. Pressurized Water Reactor (PWR) Plant
(Generic Design Features)

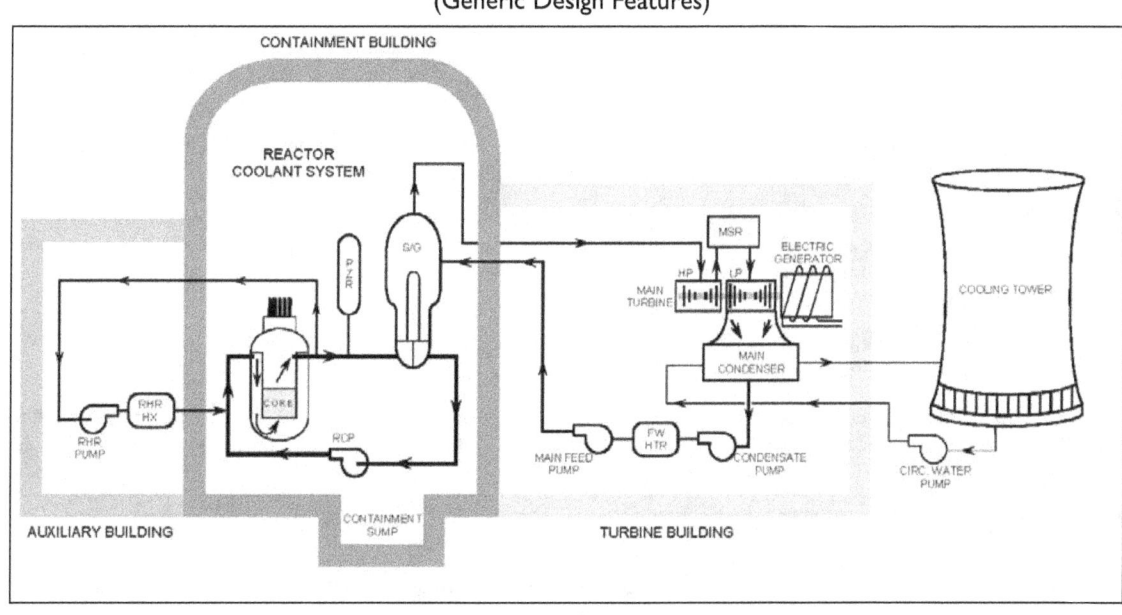

Source: U.S. Nuclear Regulatory Commission, *Reactor Concepts Manual, Boiling Water Reactor Systems, 2005.*

Notes: PIZ – Pressurizer; S/G – Steam Generator; RHR- Residual Heat Removal; RCP- Reactor Coolant Pump; HTR-Heater; MSR-Moisture Separator Reheater

To keep the reactor operating under ideal conditions, a pressurizer keeps water and steam pressure under equilibrium conditions. The pressurizer is part of the reactor coolant system, and consists of electrical heaters, pressure sprays, power-operated relief valves, and safety valves. For example, if pressure rises too high, water spray cools the steam in the pressurizer; or if pressure is too low, the heaters increase steam pressure. The cause of the pressure deviation is normally associated with a change in the temperature of the reactor coolant system.

---

[10] U.S. NRC, *Reactor Concepts Manual, Pressurized Water Reactor Systems,* http://www.nrc.gov/reading-rm/basic-ref/teachers/04.pdf - 2005-10-17.

PWR Design Evolutions

All PWR systems consist of the same major components, but arranged and designed differently. For example, Westinghouse has built plants with two, three, or four primary coolant loops, depending upon the power output of the plant.

Table 3. PWR Design Configurations

Manufacturer	Steam Generators	Reactor Coolant Pumps	Fuel Assemblies	Megawatts	Operating
Westinghouse					
Two-Loop[a]	2	2	121	500	6
Three-Loop[b]	3	3	157	700-900	13
Four Loop[c]	4	4	193	950-1,250	29
Babcock-Wilcox[d]	2	4	177	850	7
Combustion Engineering[e]	2	4		500 – 1,200	14

a. The two-loop units in the United States are Ginna, Kewaunee, Point Beach 1 and 2, and Prairie Island 1 and 2.

b. The three-loop units in the United States are Beaver Valley 1 and 2, Farley 1 and 2, H. B. Robinson 2, North Anna 1 and 2, Shearon Harris 1, V. C. Summer, Surry 1 and 2, and Turkey Point 3 and 4.

c. The four-loop units in the United States are Braidwood 1 and 2, Byron 1 and 2, Callaway, Catawba 1 and 2, Comanche Peak 1 and 2, D. C. Cook 1 and 2, Diablo Canyon 1 and 2, Indian Point 2 and 3, McGuire 1 and 2, Millstone 3, Salem 1 and 2, Seabrook, Sequoyah 1 and 2, South Texas Project 1 and 2, Vogtle 1 and 2, Watts Bar 1, and Wolf Creek.

d. The Babcock & Wilcox units in the United States are Arkansas 1, Crystal River 3, Davis Besse, Oconee 1, 2, and 3, and Three Mile Island 1.

e. The Combustion Engineering units in the United States are Arkansas 2, Calvert Cliffs 1 and 2, Fort Calhoun, Millstone 2, Palisades, Palo Verde 1, 2, and 3, San Onofre 2 and 3, Saint Lucie 1 and 2, and Waterford 3.

PWR Safe Shutdown Condition

During normal operation, a PWR does not generate steam directly. For cooling, it transfers heat via the reactor primary coolant to a secondary coolant in the steam generators. There, the secondary coolant water is boiled into steam and sent to the main turbine to generate electricity. Even after shutdown (when the moderated uranium fission is halted), the reactor continues to produce a significant amount of heat from decay of uranium fission products (decay heat). The decay heat is sufficient to cause fuel damage if the core cooling is inadequate. Auxiliary feed-water systems and the steam dump systems work together to remove the decay heat from the reactor. If a system for dumping built-up steam is not available or inoperative, atmospheric relief valves can dump the steam directly to the atmosphere. Under normal operating conditions, water flowing through the secondary system does not contact the reactor core; dumped-steam does not present a radiological release.

Loss of Coolant Accident

As with BWRs, the most severe operating condition affecting a PWR is the loss of coolant accident (LOCA); the extreme case represented by the double-ended guillotine break (DEGB) of large diameter pipe systems. In the event of a LOCA, the reactor's emergency core cooling system (ECCS) provides core cooling to minimize fuel damage by injecting large amounts of cool, borated water into the reactor coolant system from a storage tank. The borated water stops the fission process by absorbing neutrons, and thus aids in shutting down the reactor.

The ECCS on the PWR consists of four separate systems: the high-pressure injection (or charging) system, the intermediate pressure injection system, the cold leg accumulators, and the low-pressure injection system (residual heat removal). The high-pressure injection system provides water to the core during emergencies in which reactor coolant-system pressure remains relatively high (such as small breaks in the reactor coolant system, steam break accidents, and leaks of reactor coolant through a steam generator tube to the secondary side). The intermediate pressure injection system responds to emergency conditions under which the primary pressure stays relatively high; for example, small to intermediate size primary breaks. The cold leg accumulators operate without electrical power by using a pressurized nitrogen gas bubble on the top of tanks that contain large amounts of borated water. The low-pressure injection system removes residual heat by injecting water from the refueling water storage tank into the reactor coolant system during large breaks (which would cause very low reactor coolant-system pressure).

Containment Structure Designs

All U.S. reactors have primary containment structures designed to minimize releases of radioactive material into the environment. The PWR primary containment structure must surround all the components of the primary cooling system, including the reactor vessel, steam generators, and pressurizer. BWR primary containments typically are smaller, because there are no steam generators or pressurizers.

Containments must be strong enough to withstand the pressure created by large amounts of steam that the reactor cooling system may release during an accident. The largest containment designs provide sufficient space for steam released by an accident to expand and cool to keep pressure within the design parameters of the structure. Smaller containments, such as those for BWRs, require pressure suppression systems to condense much of the released steam into water. Smaller PWR containments also may include pressure suppression systems, such as ice condensers.[11]

To further limit the leakage from the containment structure following an accident, a steel liner that covers the inside surface of the containment building acts as a vapor-proof membrane to prevent any gas from escaping through any cracks that may develop in the concrete of the containment structure. Two systems act to reduce temperature and pressure within the containment structure: a fan cooler system that circulates air through heat exchangers, and a containment spray system.

All U.S. PWR designs include a containment system with multiple Engineered Safety Features (ESFs).[12] A dry containment system consists of a steel shell surrounded by a concrete biological shield that protects the reactor against outside elements, for example, debris driven by hurricane

---

[11] Kazys Almenas and R. Lee, *Nuclear Engineering: An Introduction* (Berlin: Springer-Verlag, 1992), pp. 507-514.

[12] M. Ragheb, *Containment Structures* (2011). University of Illinois Champaign-Urbana, https://netfiles.uiuc.edu/mragheb/www/NPRE%20457%20CSE%20462%20Safety%20Analysis%20of%20Nuclear%20Reactor%20Systems/Containment%20Structures.pdf.

winds or an aircraft strike.[13] The outer shield does not have a design function as a barrier against the release of radiation. Although the concrete structures in existing plants act as insulators against uncontrolled releases of radioactivity to the environment, they will fail if the ESFs fail in their function. A summary of containment building design features appears in **Table 4**.

The NRC Containment Performance Working Group studied containment buildings in 1985 to estimate their potential leak rates as a function of increasing internal pressure and temperature associated with severe accident sequences involving significant core damage.[14] It indentified potential leak paths through containment penetration assemblies (such as equipment hatches, airlocks, purge and vent valves, and electrical penetrations) and their contributions to leakage from for the containment. Because the group lacked reliable experimental data on the leakage behavior of containment penetrations and isolation barriers at pressures beyond their design conditions, it relied on an analytical approach to estimate the leakage behavior of components found in specific reference plants that approximately characterize the various containment types.

---

[13] NRC regulations require that new reactors be designed to withstand the impact of large commercial aircraft and that existing plants develop strategies to mitigate the effects of large aircraft crashes. See CRS Report RL34331, *Nuclear Power Plant Security and Vulnerabilities*, by Mark Holt and Anthony Andrews.

[14] U.S. Nuclear Regulatory Commission, *General Studies of Nuclear Reactors; BWR Type Reactors; Containment; Reactor Accidents; Leaks; PWR Type Reactors; Accidents; Reactors; Water Cooled Reactors; Water Moderated Reactors*, NUREG-1037, May 1, 1985.

Table 4. Containment Building Design Parameters

Containment Type, plant	Parameter	Technical Specification
SP-1, Zion	Containment capability pressure	149 psia[0]
	Upper bound spike pressure	107 psia
	Early failure physically unreasonable best estimate pressure rise, including heat sinks	10 psi/hour
	Time to failure, best estimate with unlimited water in cavity	16 hours
SP-2, Surry	Containment capability pressure	134 psia
	Upper bound spike pressure	107 psia
	Time to failure, early failure physically unreasonable best estimate with dry cavity	Several days
SP-3, Sequoyah	Containment capability pressure	65 psia, 330 °F
	Upper bound loading pressure	70-100 psia
	Lower bound loading pressure	50-70 psia
	Thermal loads	500-700 °F
	Early failure	Quite likely
SP-4, Browns Ferry	Containment capability pressure	132 psia, 330 °F
	Upper bound loading pressure	132 psia in 40 minutes
	Lower bound loading pressure	132 psia in 2 hours
	Thermal loads	500-700 °F
	Early failure	Quite likely
SP-6, Grand Gulf	Containment capability pressure	75 psia
	Upper bound loading pressure	30 psia
	Wall heat flux	1,000 to 10,000 Btu/hr-square foot
	Penetration seal temperature	345 °F
	Pressurization failure from diffusion flames	Unreasonable
	Seal failure	Unlikely
SP-15, Limerick	Containment capability pressure	155 psia, 330 °F
	Upper bound loading pressure	145 psia in 2-3 hours
	Lower bound loading pressure	100 psia in 3 hours
	Thermal loads	500-700 °F
	Early failure	Rather unlikely

Source: U.S. NRC, *General Studies of Nuclear Reactors; BWR Type Reactors; Containment; Reactor Accidents; Leaks; PWR Type Reactors; Accidents; Reactors; Water Cooled Reactors; Water Moderated Reactors*, NUREG-1037, 1985, as cited by M. Ragheb UICU (see footnotes).

Notes: The NRC never released NUREG-1037, but draft versions apparently circulated. Psia = pounds per square inch atmospheric.

Seismic Siting Criteria

Earthquakes occur when stresses in the earth exceed the strength of a rock mass, creating a fault or mobilizing an existing fault.[15] The fault can slip laterally (a strike/slip fault, such as the San Andreas Fault), move vertically (a thrust or reverse fault, such as the fault that caused the March 11 Japanese earthquake), or move in some combination of the two. The fault's sudden release sends seismic shock waves through the earth that have two primary characteristics: (1) amplitude—a measure of the peak wave height, and (2) period—the time interval between the arrival of successive peaks or valleys.[16] The seismic wave's arrival causes ground motion. The intensity of ground motion depends primarily on three factors: the distance from the source (also known as focus or epicenter), the amount of energy released (magnitude of the earthquake), and the type of soil or rock at the site.

In general, for a given magnitude earthquake, the shallower the focus, the stronger the wave will be when reaching the surface. In addition, the intensity of ground shaking diminishes with increasing distance from the earthquake focus. Sites with deep, soft soils or loosely compacted fill will experience stronger ground motion than sites with stiff soils or rock.

> **Earthquake Magnitude**
>
> The common measure of an earthquake's magnitude (*M*) refers to the logarithmic Richter scale, thus an *M* 7.0 earthquake has an amplitude that is ten times larger than an *M* 6.0, but releases 31.5 times more energy than an *M* 6.0 earthquake.

Safe Shutdown Earthquake Condition

In 1973, the concept of the "safe shutdown earthquake" (SSE) was introduced in Title 10 Part 100 of the *Code of Federal Regulations* (10 CFR 100), *Appendix A—Seismic and Geologic Siting Criteria for Nuclear Power Plants*. The NRC defines the Safe Shutdown Earthquake as the maximum earthquake in which certain structures, systems, and components, important to safety, must remain functional.[17] Under an "operating basis earthquake," the reactor could continue operation without undue risk to the safety of the public.

Ground motion at any specific location, such as a nuclear plant site, depends on the earthquake source, magnitude, distance to the source, and the attenuation (dampening) caused by rock and soil characteristics. A nuclear power plant responds to an earthquake depending on how its individual structures, systems, and components resonate, or vibrate, with the ground shaking. Heavier and more massive structures resonate at lower frequencies, while light components resonate at higher frequencies.

During an earthquake, ground motion transmits vibrations to a nuclear power plant's foundation and structure. The vibrations cause back-and-forth acceleration of a structure, system or components that is measured relative to the earth's gravitational acceleration constant (g). Both

---

[15] The Applied Technology Council (ATC) and the Structural Engineers Association of California (SEAOC), *Briefing Paper 1 Building Safety and Earthquakes Part A: Earthquake Shaking and Building Response*, Redwood City, CA, http://www.atcouncil.org/.

[16] The wave's frequency is the inverse of the period ($1/s$), and is expressed as the number of wave cycles per second (termed Hertz or *Hz*).

[17] http://www.nrc.gov/reading-rm/basic-ref/glossary/safe-shutdown-earthquake.html

---

vertical and horizontal components of ground acceleration place loads, or stresses, on a nuclear power plant's structure.[18] Peak Ground Acceleration (PGA) is a measure that has been widely used in developing nuclear power plant "fragility estimates," which represent the sensitivity of nuclear plant structures, systems, and components (SSCs) to the inertial effects of acceleration during ground shaking.

Cumulative Absolute Velocity

Structural damage to nuclear power plants occurs when the cumulative effects of ground acceleration (seismically induced vibrations) cross a certain threshold. The Electric Power Research Institute (EPRI) developed the concept of "cumulative absolute velocity" (CAV) in 1988 as an index for indicating the onset of structural damage from the cumulative effects of ground acceleration.[19] The threshold between damaging and non-damaging earthquakes (for well-designed buildings) conservatively occurs at ground motions with cumulative absolute velocities (CAV) greater than 0.16 g-seconds.[20] In simple terms, CAV is the sum of various ground acceleration frequencies (measured in terms of g) and the duration of their acceleration (measured in seconds). An example of this phenomenon is a wire coat-hanger that breaks from metal fatigue after being rapidly bent multiple times.

Experimental and empirical seismic data have provided insights into the behavior of different structures under various acceleration and shaking conditions. For example, welded steel piping at nuclear power plants rarely failed when peak ground accelerations remained below 0.5g.[21] Other types of structures exhibit different behaviors. Engineers design the various plant structures to withstand a certain severity of earthquake and estimates of ground shaking specific to each plant site.

The maximum vibratory accelerations of the Safe Shutdown Earthquake must take into account the characteristics of the underlying soil material in transmitting the earthquake-induced motions at the various locations of the plant's foundation. Various plant structures, depending upon their elevation above the foundation, vibrate at different frequencies during an earthquake. Vibrations in the range of 1 to 10 Hz are particularly problematic, because a wide range of structures are susceptible to damaging resonance at those frequencies.[22] These accelerations and the corresponding shaking frequencies are factors in the Probabilistic Seismic Hazard Analysis (PSHA, discussed below). The full seismic spectrum often can be characterized by two intervals: peak ground acceleration (PGA) and spectral acceleration (SA) averaged between 5 and 10 hertz (Hz).

---

[18] Gravitation acceleration g = 32 feet/second/second (ft/second^2).

[19] Kenneth W. Campbell and Yousef Bozorgnia, "A Ground Motion Prediction Equation for the Horizontal Component of Cumulative Absolute Velocity on the PEER-NGA Strong Motion Database," *Earthquake Spectra*, vol. 26, no. 3 (August 2010), p. 635.

[20] Nuclear Regulatory Commission, *A Performance-Based Approach to Define the Site Specific Earthquake Ground Motion*, regulatory Guide 1.208, March 2007, p. 7.

[21] N.C. Chokshi, S.K. Shaukat, and A.L. Hiser, et al., *Seismic Considerations for the Transition Break Size*, U.S. NRC, NUREG-1903, February 2008, pp. 29-30.

[22] Frequency *Hz* (Hertz) refers to the number of cycles per second (which is inverse of the ground motion wave period — the time between two wave peaks). Thus, 0.2-*s* is the equivalent of 5 *Hz* (1/0.2-*s*), and 1-*s* is the equivalent of 1 Hz (1/1-*s*).

Seismic Design Varies by Region

In the western United States (WUS), where earthquakes with frequencies below 15 Hz predominate, earthquake magnitude is a principal design consideration for nuclear power plants.[23] Earthquakes below the 10 Hz frequency range pose the greatest hazard to nuclear power plants.[24] In the central and eastern United States (CEUS), designs considered both earthquake magnitude and Modified Mercalli Intensity (MMI) due to sparse recordings of actual earthquake events.[25] While plants designed to operate in the CEUS must also withstand low frequency earthquakes, the earthquakes that do occur are associated more often with higher frequencies than in the WUS. Higher frequency earthquakes are less damaging to large structures but may adversely affect small components.[26]

Deterministic Seismic Hazard Analysis

By the late 1940s, structural engineers had begun considering the shear forces caused by earthquakes that structures must resist. To supplement their design calculations, they referred to the Seismic Zone Map published by the Uniform Building Code (UBC).[27] (Refer to Appendix **Figure B-1**.) The UBC map divided the United State into six distinct seismic zones representing various degrees of seismic risk. The map expressed peak ground acceleration as the decimal ratio of the gravitational acceleration constant (g) that applied to a Maximum Credible Earthquake (MCE) and an Operating Basis Earthquake (OBE). UBC defined a maximum credible earthquake and its associated ground motion as the largest magnitude earthquake that could reasonably occur along the recognized faults or within a particular seismic source. An operating basis earthquake was defined as having the greatest level of ground motion likely to occur during the economic life of a structure.

Designs for nuclear power plants granted construction permits during the 1960s and 1970s applied a deterministic approach to seismic design based on site-specific investigations of local and regional seismology, geology, and geotechnical soil conditions to determine the maximum credible earthquake from a single source (fault).[28] Deterministic Seismic Hazard Analysis (DSHA) attempted to quantify the effects of a maximum credible earthquake based on known seismic sources sufficiently near the site and available historical seismic and geological data to estimate ground motion at the plant site.[29]

Appendix A to 10 CFR 100 requires an investigation of fault and earthquake occurrences to provide the basis for determining a safe shutdown earthquake. Appendix A to 10 CFR 100 notes

---

[23] See **Appendix A** for a discussion of earthquake magnitude.

[24] J. Hamel, K. Huffman, and R. Kassawara, "Nuclear Seismic Safety: Modeling Risk in the Real World," *EPRI Journal*, Summer 2010, p. 15.

[25] However, in the CEUS, magnitude is increasingly used as the measure of earthquake size, and ground motions are correspondingly estimated using correlations with magnitude.

[26] Hamel et al.

[27] The International Building Code (IBC) published by the International Code Council (ICC) replaced the UBC in 2000.

[28] U.S. Nuclear Regulatory Commission, *Evaluation of the Seismic Design Criteria in ASCE/SEI Standard 43-05 for Application to Nuclear Power Plants*, NUREG/CR-6926, Brookhaven National Laboratory, NY, March 2007.

[29] U.S. Army Corps of Engineers, *Earthquake Design and Evaluation for Civil Works Projects* (ER 1110-2-1806), July 31, 1995.

---

the limitations for basing seismic design criteria on literature reviews of geophysical and geologic information, and requires supplementing the investigation with studies for vibratory ground motion, evidence of surface faulting, and evidence of seismically induced floods and water waves that have or could have affected the site.

Probabilistic Seismic Hazard Analysis

Under 10 CFR 100.23 (Geologic and Seismic Siting Criteria), designs for new nuclear power plants will base their Safe Shutdown Earthquake on Probabilistic Seismic Hazard Analysis (PSHA). The methodology has also found widespread use in U.S. engineering practice for non-nuclear structures. Where DSHA had based peak ground acceleration (PGA) on a single earthquake source, PSHA uses up-to-date interpretations of earthquake sources, earthquake recurrence, and strong ground motion estimates to estimate the probability of exceeding various levels of earthquake-caused ground motion at a given location in a given future time period.[30] It quantifies a site's seismic hazard characteristics from seismic hazard curves or "response spectra" developed in part by identifying and characterizing each seismic source in terms of maximum magnitude, magnitude recurrence relationship, and source geometry.

A response spectrum is a plot of the maximum response (acceleration, velocity, or displacement) of a family of oscillations (ground or structures). When derived from a earthquake record, the site-specific ground motion response spectrum appears as an irregular graph of peaks and valleys that combines a number of individual response spectra from past earthquakes (**Figure 7**).

---

[30] R. J. Budnitz, G. Apostolakis, and D. M. Boore, *Recommendations for Probabilistic Seismic Hazard Analysis: Guidance on Uncertainty and Use of Experts: Main Report*, U.S. Nuclear Regulatory Commission, Nureg/CR-6372, Lawrence Berkeley National Laboratory, CA, April 1997, http://www.nrc.gov/reading-rm/doc-collections/nuregs/contract/cr6372/vol1/index html#pub-info.

Figure 7. Constructing Site-Specific Ground Motion Response Spectrum

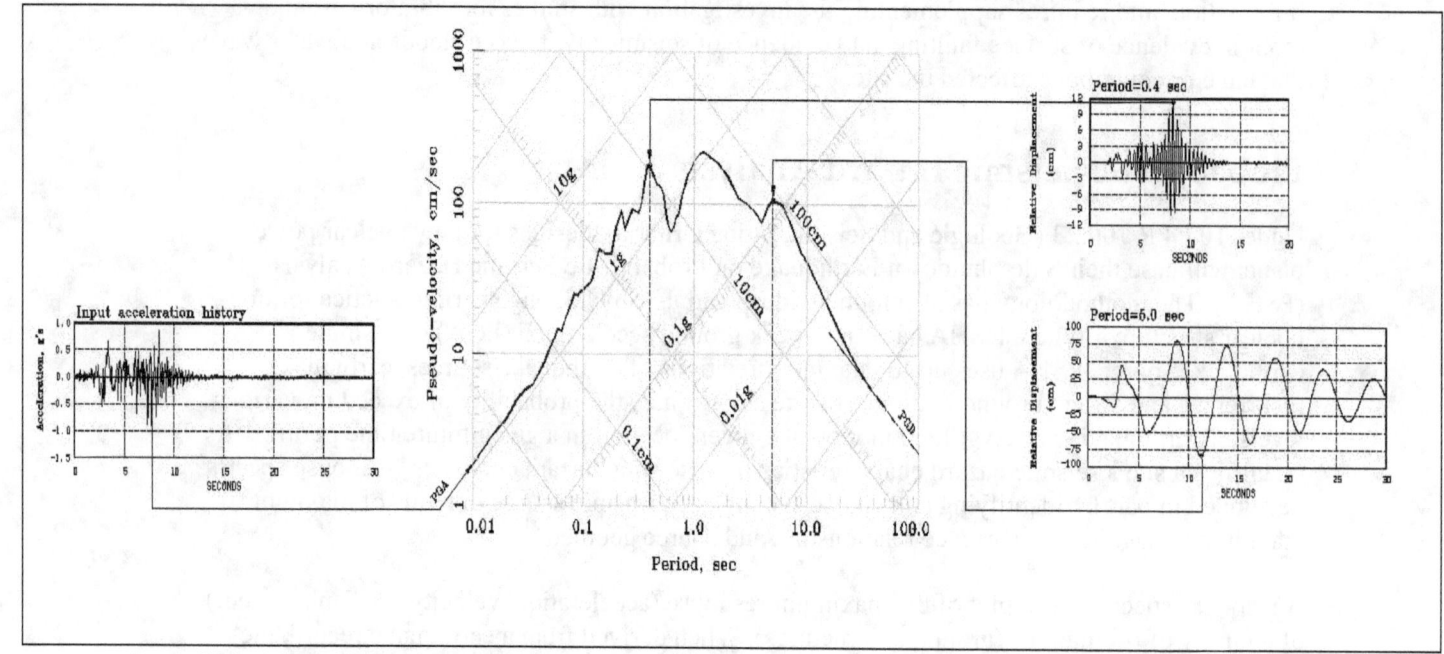

Source: U.S Army Corps of Engineers Response Spectra and Seismic Analysis for Concrete Hydraulic Structures EM 1110-2-6050, June 30, 1999.

Notes: Each earthquake produces a unique sequence of ground motions (accelerations) that may last several seconds or longer. The record of ground motion, captured on an accelerograph, appears as a jagged-shaped line that represents the peak values of acceleration/de-acceleration. The ground motion response spectrum represents the range of multiple earthquake records.

CRS-20

Design Response Spectra for Seismic Design of Nuclear Power Plants

Each earthquake produces a spectrum of ground motions that vary in frequency and acceleration. The seismic spectra important to nuclear power plant design are peak ground accelerations between 5 and 10 Hz. The NRC has developed Design Response Spectra statistically from response spectra of past strong motion earthquakes. The former Atomic Energy Commission (AEC) (the NRC's predecessor) published Regulatory Guide 1.60, *Design Response Spectra of Nuclear Power Reactors* in 1973 to provide spectral shapes for horizontal and vertical ground movements that designs must respond to (design response).

A Safe Shutdown Earthquake is defined by 10 CFR 100 Appendix A as the response spectra corresponding to the maximum vibratory accelerations at the elevations of the nuclear power plant structural foundations. NRC may credit nuclear power plant foundations with a 5% dampening affect in reducing the transmission of ground accelerations. In the example of **Figure 8**, the range of maximum accelerations (g) is plotted against the range of corresponding frequencies (Hz) for the Surry Nuclear Power Plant containment building.

Figure 8. NRC Site Seismic Design Response Spectra

(U.S. NRC Regulatory Guide 1.60 Design Response Spectra—5% damping)

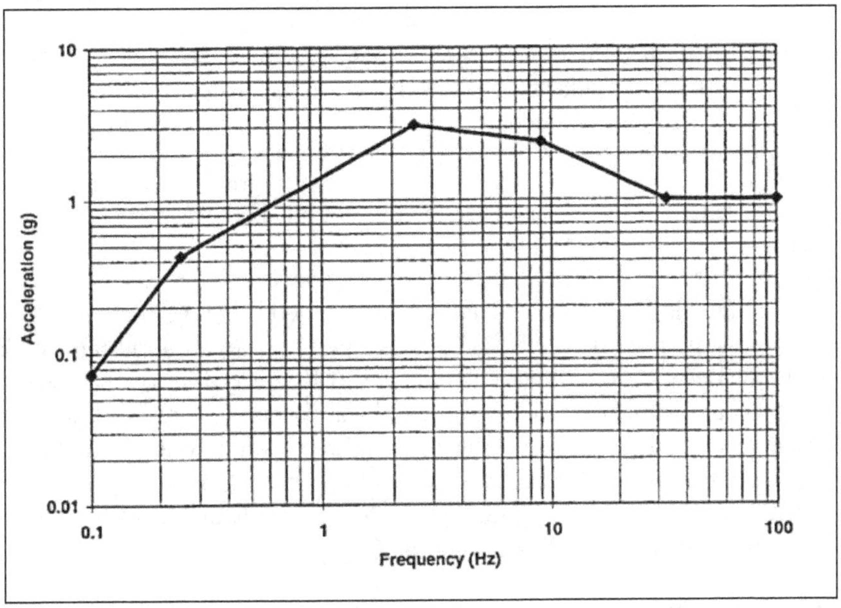

Source: U.S. NRC, Structural Seismic Fragility Analysis of the Surry Containment, Figure 3.1 NUREG/CR-6783, June 2002.

Notes: The seismic spectrum important to nuclear power plant design is characterized by two intervals—peak ground acceleration and spectral acceleration averaged between 5 and 10 Hz. The NRC considers that plant foundations reduce the transmission of ground accelerations and credits the foundations with a 5% dampening affect.

The NRC requires that nuclear plant designs account for site-specific ground motions and has specified a minimum ground motion level for nuclear plant designs. The NRC Regulatory Guide

1.208 endorses either the EPRI or Lawrence Livermore National Laboratory (LLNL) seismic hazard models as a starting point for conducting a PSHA in siting nuclear power plants.

In 2007, the NRC contracted with Brookhaven National Laboratory to produce *Evaluation of the Seismic Design Criteria in ASCE/SEI Standard 43-05 for Application to Nuclear Power Plants* (NUREG/CR-6926). The report presents the results of a review of the American Society of Civil Engineers/Structural Engineering Institute (ASCE/SEI) Standard 43-05, "Seismic Design Criteria for Structures, Systems, and Components in Nuclear Facilities." As its title implies, this standard provides seismic design criteria for safety-related structures, systems, and components (SSCs) in a broad spectrum of nuclear facilities.

National Seismic Hazard Maps

In 2008, the U.S. Geological Survey (USGS) released an update of the National Seismic Hazard Maps (NSHM).[31] The purpose of the maps is to show the likelihood of a particular severity of shaking within a specified time-period. The Seismic Hazard maps are the basis for seismic design provisions of building codes to allow buildings, highways, and critical infrastructure to withstand earthquake shaking without collapse.

The USGS revises the NHSM every six years to reflect newly published earthquake data. The NHSM are used to update building code seismic design provisions. USGS notes that the 2008 hazard maps differ significantly from the 2002 maps in many parts of the United States:

> The new maps generally show 10- to 15-percent reductions in acceleration across much of the Central and Eastern United States [CEUS] for 0.2-s [second] and 1.0-s spectral acceleration and peak horizontal ground acceleration for 2-percent probability of exceedance in 50 years. The new maps for the Western United States [WUS] indicate about 10-percent reductions for 0.2-s spectral acceleration and peak horizontal ground acceleration and up to 30-percent reductions in 1.0-s spectral acceleration at similar hazard levels.[32]

Although the seismic hazard is highest in the most tectonically active regions of the western United States, including Alaska, the central and eastern United States (CEUS) also contain regions of elevated seismic hazard. The USGS has mapped the New Madrid Seismic Zone (NMSZ)—which includes parts of Arkansas, Tennessee, Missouri, Illinois, and Kentucky—as comparable to the more seismically active portions of California. Portions of Indiana are also subject to elevated seismic hazard from the Wabash Valley Seismic Zone along the border of Illinois and Indiana. Earthquakes also occur in the central United States outside of these two documented zones, however, as demonstrated by the November 6, 2011, magnitude 5.6 earthquake that occurred 40 miles east of Oklahoma City.

In the CEUS, the New Madrid Seismic Zone and the Charleston area in southeast South Carolina comprise the dominant seismic hazard (at 2% probability of exceedance in 50 years). Seismically active portions of eastern Tennessee and some portions of the northeast also contribute to the seismic hazard. The hazard at the 2% probability of exceedance in 50 years level is typically a

---

[31] Mark D. Petersen, Arthur D. Frankel, and Stephen C. Harmsen et al., *Documentation for the 2008 Update of the United States National Seismic Hazard Maps*, U.S Geological Survey, Open-File Report 2008 1128, 2008, http://earthquake.usgs.gov/hazards/

[32] Ibid.

factor of two to four times higher than the 10% probability of exceedance in 50 years values in the seismically active portions of the CEUS. The elevated seismic hazard in the New Madrid Seismic Zone is due primarily to three large earthquakes of magnitude 7.0 or greater that occurred during 1811-1812 and the chances of an earthquake of similar magnitude striking the region again. Because of the elevated hazard, portions of the NMSZ in the central United States require monitoring by a relatively dense array of seismometers compared to the northeastern and southeastern United States. As discussed earlier, most of the nation's nuclear power plants are located in the CEUS.

The dominant earthquake hazard in the northeastern United States is related to seismic activity in the St. Lawrence rift zone, which extends from northeastern New York State into Canada. However, other parts of the northeast have also experienced earthquakes, including western New York, Massachusetts, New Jersey, Pennsylvania, Ohio, and Maine.[33]

The seismic hazard in the southeastern United States is primarily near Charleston, South Carolina, and along the Eastern Tennessee Seismic Zone. The largest historically documented earthquake that occurred in the CEUS other than the 1811-1812 earthquakes in the NMSZ took place near Charleston in 1886. The Eastern Tennessee Seismic Zone comprises a swath including portions of Alabama, Georgia, Tennessee, North Carolina, and Virginia. The magnitude 5.8 earthquake that struck Virginia on August 23, 2011, may have been in the northernmost portion of the Eastern Tennessee Seismic Zone.

Figure 9 and **Figure 10** show the locations of nuclear plant sites on the 2008 USGS National Seismic Hazard Map for the United States. These maps display quantitative information about seismic ground motion hazards expressed as horizontal ground acceleration (g) of a particle at ground level moving horizontally during an earthquake. CRS cautions against drawing any conclusion regarding a plant's seismic risk from the figures. **Figure 11** shows the proximity of plant sites to Quaternary period faults based on the USGS Quaternary Fault and Fold Database of the United States.[34] (Quaternary-active faults are those that have slipped in Quaternary time—the last 1.6 million years.) The map is not a prediction of an earthquake event. The USGS Database has information on faults and associated folds in the United States that are believed to be sources of greater than magnitude 6 earthquakes during the Quaternary period. Geologists think that these faults are the most likely source of future great earthquakes, so it is important to know what they are, where they are, and how they work.[35]

---

[33] The largest earthquakes in New York, New Jersey, and Massachusetts were, respectively: 1944, Massena, NY, magnitude 5.8, felt from Canada south to Maryland; 1783, New Jersey, magnitude 5.3, felt from New Hampshire to Pennsylvania; and 1755, Cape Ann and Boston, MA, intensity of VIII on the Modified Mercalli Scale, felt from Nova Scotia to Chesapeake Bay (USGS Earthquake Hazards Program).

[34] The Quaternary Period of the geologic time scale encompasses the past 1.6 million years.

[35] U.S.G.S, *Quaternary Map Introduction*, http://geomaps.wr.usgs.gov/sfgeo/quaternary/.

Figure 9. Operating Nuclear Power Plant Sites vs. Seismic Hazard

(Seismic hazard expressed as horizontal ground acceleration in terms of percent of the gravitation acceleration constant)

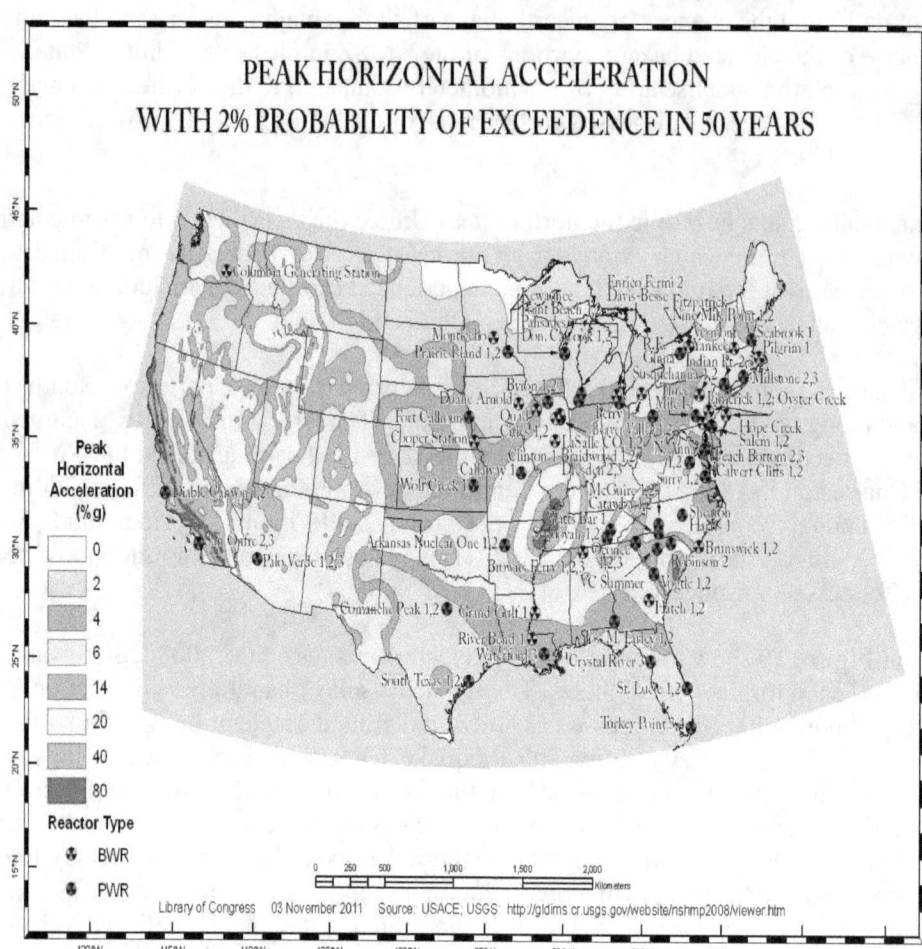

Source: USGS Seismic Hazard Map for the United States, http://earthquake.usgs.gov/hazards/products/conterminous/2008/maps/, prepared for CRS by the Library of Congress Geography and Maps Division.

Notes: This map displays quantitative information about seismic ground motion hazards as horizontal ground acceleration (in terms of gravitational acceleration) of a particle at ground level moving horizontally during an earthquake. This map is not a prediction of an earthquake event. The NRC does not rank nuclear plants by seismic risk. No commercial nuclear power plants operate in either Alaska or Hawaii.

Figure 10. Operating Nuclear Power Plants vs. Seismic Hazard

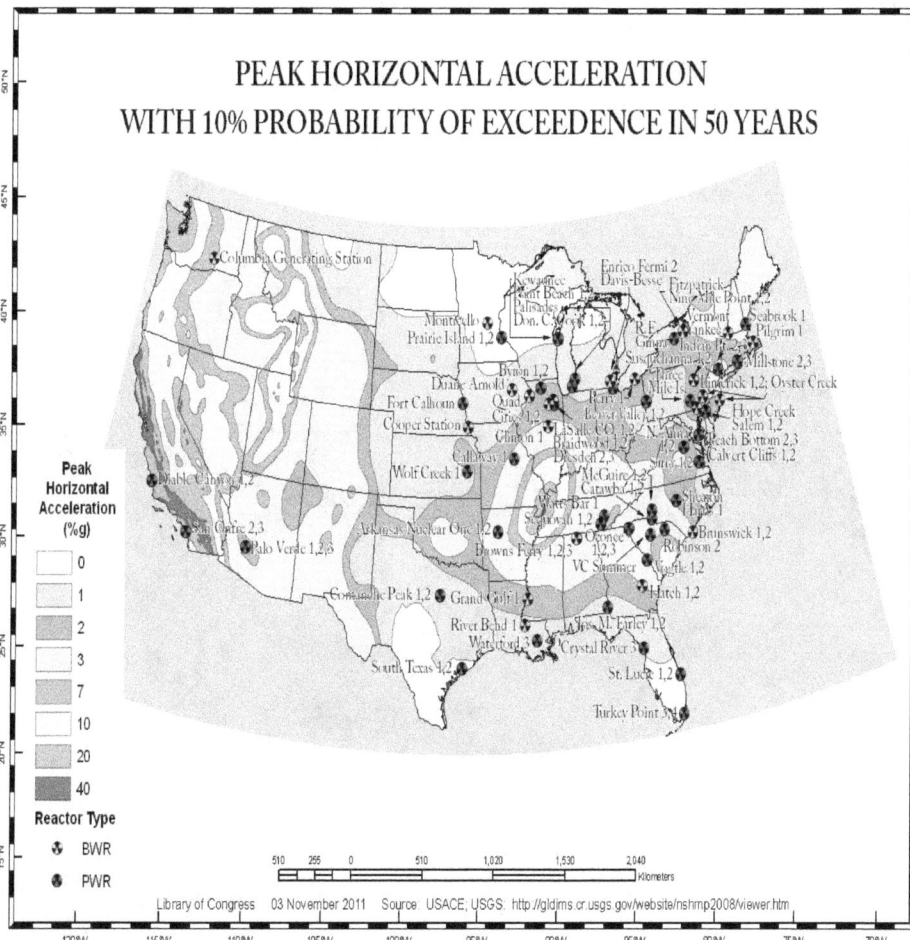

Source: USGS Seismic Hazard Map for the United States, http://earthquake.usgs.gov/hazards/products/conterminous/2008/maps/ prepared for CRS by the Library of Congress Geography and Maps Division.

Notes: This map displays quantitative information about seismic ground motion hazards as horizontal ground acceleration (in terms of gravitational acceleration) of a particle at ground level moving horizontally during an earthquake. This map is not a prediction of an earthquake event. The NRC does not rank nuclear plants by seismic risk. No commercial nuclear power plants operate in either Alaska or Hawaii.

Figure 11. Operating Nuclear Power Plant Sites and Mapped Quaternary Faults

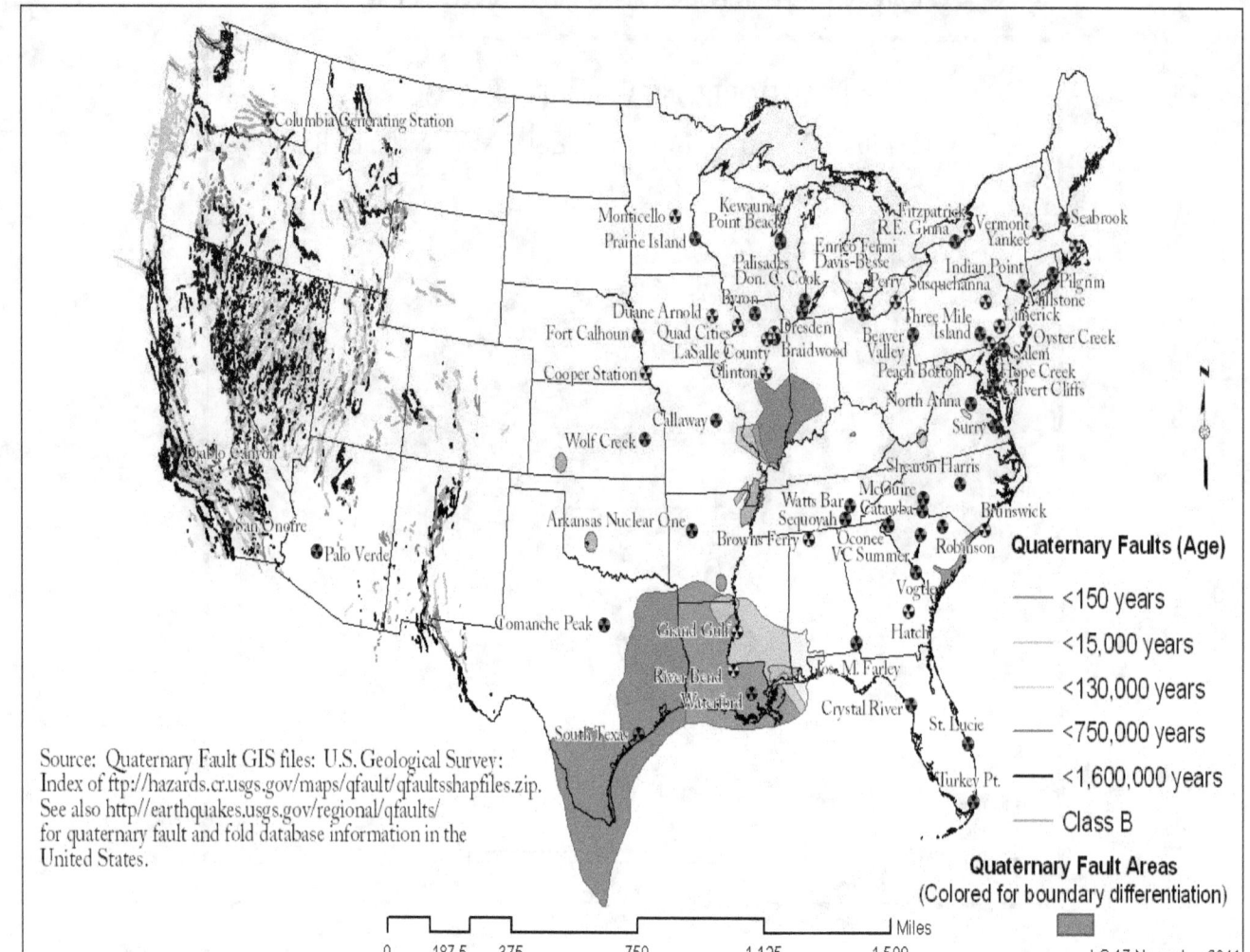

Source: CRS and the USGS Quaternary Fault and Fold Database of the United States.

Notes: Indicates tectonic sources of greater than moment magnitude 6 (M>6) earthquakes during the Quaternary, except for growth faults in the Gulf Coast. Quaternary-Faults are those that have slipped in Quaternary time (the last 1.6 million years). Class B faults are thought to be not capable of producing earthquakes and are older than the Quaternary in age. This map is not a prediction of an earthquake event. No commercial nuclear power plants operate in either Alaska or Hawaii.

Notes:

Reactor Name	State	Type	Megawatts	Operating License Issued
Arkansas Nuclear One 1	AR	PWR	843	1974
Arkansas Nuclear One 2	AR	PWR	995	1978
Beaver Valley 1	PA	PWR	892	1976
Beaver Valley 2	PA	PWR	846	1976
Braidwood 1	IL	PWR	1178	1987
Braidwood 2	IL	PWR	1152	1988
Browns Ferry 1	AL	BWR	1065	1973
Browns Ferry 2	AL	BWR	1104	1974
Browns Ferry 3	AL	BWR	1115	1976
Brunswick 1	NC	BWR	938	1976
Brunswick 2	NC	BWR	937	1974
Byron 1	IL	PWR	1164	1985
Byron 2	IL	PWR	1136	1987
Callaway 1	MO	PWR	1236	1984
Calvert Cliffs 1	MD	PWR	873	1974
Calvert Cliffs 2	MD	PWR	862	1976
Catawba 1	SC	PWR	1129	1985
Catawba 2	SC	PWR	1129	1986
Clinton 1	IL	BWR	1065	1987
Columbia Generating Station	WA	BWR	1190	1984
Comanche Peak 1	TX	PWR	1200	1990
Comanche Peak 2	TX	PWR	1150	1993
Cooper Station	NE	BWR	830	1974
Crystal River 3	FL	PWR	838	1976
Davis-Besse	OH	PWR	893	1977
Diablo Canyon 1	CA	PWR	1151	1984
Diablo Canyon 2	CA	PWR	1149	1985
Donald C. Cook 1	MI	PWR	1009	1974
Donald C. Cook 2	MI	PWR	1060	1977
Dresden 2	IL	BWR	867	1991
Dresden 3	IL	BWR	867	1971
Duane Arnold	IA	BWR	640	1974
Enrico Fermi 2	MI	BWR	1122	1985
Fitzpatrick	NY	BWR	852	1974
Fort Calhoun	NE	PWR	500	1973
Grand Gulf 1	MS	BWR	1297	1984
Hatch 1	GA	BWR	876	1974
Hatch 2	GA	BWR	883	1978
Robinson 2	SC	PWR	710	1971
Hope Creek 1	NJ	BWR	1061	1986
Indian Point 2	NY	PWR	1023	1973
Indian Point 3	NY	PWR	1025	1975
Joseph M. Farley 1	AL	PWR	851	1977
Joseph M. Farley 2	AL	PWR	860	1981
Kewaunee	WI	PWR	556	1973
LaSalle County 1	IL	BWR	1118	1982
LaSalle County 2	IL	BWR	1120	1983
Limerick 1	PA	BWR	1134	1985
Limerick 2	PA	BWR	1134	1989
McGuire 1	NC	PWR	1100	1981
McGuire 2	NC	PWR	1100	1983
Millstone 2	CT	PWR	884	1975
Millstone 3	CT	PWR	1227	1986
Monticello	MN	BWR	579	1970
Nine Mile Point 1	NY	BWR	621	1974
Nine Mile Point 2	NY	BWR	1140	1987
North Anna 1	VA	PWR	981	1978

Reactor Name	State	Type	Megawatts	Operating License Issued
North Anna 2	VA	PWR	973	1980
Oconee 1	SC	PWR	846	1973
Oconee 2	SC	PWR	846	1973
Oconee 3	SC	PWR	846	1974
Oyster Creek	NJ	BWR	619	1991
Palisades	MI	PWR	778	1971
Palo Verde 1	AZ	PWR	1335	1985
Palo Verde 2	AZ	PWR	1335	1986
Palo Verde 3	AZ	PWR	1335	1987
Peach Bottom 2	PA	BWR	1112	1973
Peach Bottom 3	PA	BWR	1112	1974
Perry 1	OH	BWR	1261	1986
Pilgrim 1	MA	BWR	685	1972
Point Beach 1	WI	PWR	512	1970
Point Beach 2	WI	PWR	514	1973
Prairie Island 1	MN	PWR	551	1974
Prairie Island 2	MN	PWR	545	1974
Quad Cities 1	IL	BWR	867	1972
Quad Cities 2	IL	BWR	869	1972
R. E. Ginna	NY	PWR	498	1969
River Bend 1	LA	BWR	989	1985
Salem 1	NJ	PWR	1174	1976
Salem 2	NJ	PWR	1130	1981
San Onofre 2	CA	PWR	1070	1982
San Onofre 3	CA	PWR	1080	1992
Seabrook 1	NH	PWR	1295	1990
Sequoyah 1	TN	PWR	1148	1980
Sequoyah 2	TN	PWR	1126	1981
Shearon Harris 1	NC	PWR	900	1986
South Texas 1	TX	PWR	1410	1988
South Texas 2	TX	PWR	1410	1989
St. Lucie 1	FL	PWR	839	1976
St. Lucie 2	FL	PWR	839	1983
Surry 1	VA	PWR	799	1972
Surry 2	VA	PWR	799	1973
Susquehanna 1	PA	BWR	1149	1982
Susquehanna 2	PA	BWR	1140	1984
Three Mile Island 1	PA	PWR	786	1974
Turkey Point 3	FL	PWR	720	1972
Turkey Point 4	FL	PWR	720	1973
VC Summer	SC	PWR	966	1982
Vermont Yankee	VT	BWR	510	1972
Vogtle 1	GA	PWR	1109	1987
Vogtle 2	GA	PWR	1127	1989
Waterford 3	LA	PWR	1250	1985
Watts Bar 1	TN	PWR	1123	1996
Wolf Creek 1	KS	PWR	1166	1985

NRC Review—Implications of Updated Probabilistic Seismic Hazard Estimates in Central and Eastern United States on Existing Plants

The NRC does not rank nuclear plants by seismic risk, and the NRC has not published a regulatory guide that recommends using the USGS national hazard maps for siting nuclear power plants. However, it recently considered the implications of updated USGS seismic hazard models on the seismic risk of nuclear power plants sites operating in the Central and Eastern United states. The NRC has required that each nuclear plant built meet certain structural specifications based on the earthquake susceptibility of each plant site. The NRC may re-evaluate some of those design specifications in light of the 2008 USGS seismic hazard maps. In 2010, the NRC published *Review Implications of Updated Probabilistic Seismic Hazard Estimates in the Central and Eastern United States on Existing Plants* (GI-199 Safety/Risk Assessment), a two-stage assessment that determines the implications of the 2008 USGS updated probabilistic seismic hazard maps in the CEUS on existing nuclear power plant sites.[36] NRC's objective in the GI-199 Safety/Risk Assessment was to evaluate the need for further investigations of seismic safety for operating reactors in the CEUS.

The assessment first evaluated the change in seismic hazard with respect to previous estimates at individual nuclear power plants, and then estimated the change in Seismic Core Damage Frequency (SCDF) resulting from the change in the seismic hazard. Seismic core damage frequency is the probability of damage to the reactor core (fuel rods) resulting from a seismic initiating event. It does not necessarily imply that a core meltdown or loss of containment (associated with a radiological release) would occur. The seismic hazard at each plant site depends on the unique seismology and geology surrounding the site. Consequently, the report separately determined the implications of updated probabilistic seismic hazard for each of the 96 operating nuclear power plants in the CEUS.[37]

The data evaluated in the assessment suggest that the probability for earthquake ground motion above the seismic design basis for some nuclear plants in the CEUS, although still low, is larger than previous estimates. In March 2011, the NRC announced that it had identified 27 nuclear reactors operating in the CEUS (listed in **Table 5**) subject to priority earthquake safety reviews.[38]

Table 5. Operating Nuclear Power Plants Subject to Earthquake Safety Reviews

Plant	St.	Type	Plant	St.	Type	Plant	St.	Type
Crystal River 3	FL	PWR	North Anna 1 & 2	VA	PWR	Sequoyah 1 & 2	TN	PWR
Dresden 2 & 3	IL	BWR	Oconee 1, 2 & 3	SC	PWR	Seabrook	NH	PWR
Duane Arnold	IA	BWR	Perry 1	OH	BWR	V.C. Summer	SC	PWR
Joseph M. Farley 1 & 2	AL	PWR	Peach Bottom 2 & 3	PA	BWR	Watts Bar 1	TN	PWR
Indian Point 2 & 3	NY	PWR	River Bend 1	LA	BWR	Wolf Creek	KS	PWR
Limerick 1 & 2	PA	BWR	Saint Lucie 1 & 2	FL	PWR			

Source: The Energy Daily.

Note: The NRC has not announced a schedule for completing the seismic reviews at the time of this report.

Recent Legislative Activities

On March 17, 2011, the Senate Committee on Homeland Security and Governmental Affairs held a hearing on Catastrophic Preparedness that looked at technologies and emergency procedures

---

[36] U.S. Nuclear Regulatory Commission, *Implications of Updated Probabilistic Seismic Hazard Estimates in Central and Eastern United States Existing Plants—Safety/Risk Assessment*, Generic Issue 199 (GI-199), August 2010.

[37] Ibid.

[38] George Lobsenz, "NRC Task Force to Review Safety: 27 Reactors Are Seismic Priorities," *The Energy Daily*, March 24, 2011.

used in the event of a large-scale earthquake or other natural disaster.[39] On April 6, 2011, the Subcommittee on Oversight and Investigations of the House Energy and Commerce Committee held a hearing on the U.S. Government Response to the Nuclear Power Plant Incident in Japan.[40] On April 7, 2011, the Subcommittee on Technology and Innovation of the House Science, Space, and Technology Committee held a hearing on Earthquake Risk Reduction.[41]

Legislation introduced in the Senate and House would reauthorize appropriations and make some changes to the ongoing National Earthquake Hazards Reduction Program (NEHRP). H.R. 1379 and S. 646 would amend the Earthquake Hazards Reduction Act of 1977 (42 U.S.C. 7704), and reauthorize appropriations through FY2015 (authorization for appropriations for NEHRP activities ended in FY2009). H.R. 3479 would also make some changes to NEHRP, and authorize appropriations through FY2014. Activities conducted under NEHRP are intended to improve the scientific understanding of earthquakes and their effects on people and infrastructure, develop effective measures to reduce earthquake hazards, and promote the adoption of earthquake hazard reduction activities. Results from these activities could improve the current understanding of how earthquake-caused shaking would affect nuclear power plants.

H.R. 1268, the Nuclear Power Licensing Reform Act of 2011, would amend Section 103 of the Atomic Energy Act of 1954 (42 U.S.C. 2133), subsection c, by adding at the end the following requirements for nuclear power plant licenses:

> Any such renewal shall be subject to the same criteria and requirements that would be applicable for an original application for initial construction, and the Commission shall ensure that any changes in the size or distribution of the surrounding population, or seismic or other scientific data not available at time of original licensing, have not resulted in the facility being located at a site at which a new facility would not be allowed to be built.

H.R. 1242, the Nuclear Power Safety Act of 2011, would amend the Atomic Energy Act to revise regulations to ensure that nuclear facilities licensed under the act can withstand and adequately respond to an earthquake, tsunami (for a facility located in a coastal area), strong storm, or other event that threatens a major impact to the facility; a loss of the primary operating power source for at least 14 days; and a loss of the primary backup operating power source for at least 72 hours.

---

[39] Senate Committee on Homeland Security and Governmental Affairs, Catastrophic Preparedness: How Ready is FEMA for the Next Big Disaster? http://hsgac.senate.gov/public/index.cfm?FuseAction=Hearings.Hearing& Hearing_ID=a42880b1-22fc-4890-b82c-dd2a369e2aa2.

[40] House Energy and Commerce Committee, The U.S. Government Response to the Nuclear Power Plant Incident in Japan, http://energycommerce.house.gov/hearings/hearingdetail.aspx?NewsID=8420.

[41] House Committee on Science, Space, and Technology, Subcommittee Reviews Status of U.S. Earthquake Preparedness, http://science.house.gov/press-release/subcommittee-reviews-status-us-earthquake-preparedness.

Policy Considerations for Monitoring Earthquakes in the CEUS in Support of Seismic Assessments of Nuclear Power Plants

The USGS developed a 2011 report in response to a request from the NRC evaluating seismic monitoring capabilities in the CEUS, with particular emphasis on meeting the current and future needs of the NRC.[42] The USGS report recommends adding 100 new stations to existing arrays of seismometers in the CEUS:

- 38 deployed in seismically active zones in the CEUS to capture information from infrequent earthquakes for use in improving the understanding of how, where, and why earthquakes occur in the region; and

- 62 stations deployed near nuclear power plants to understand the site-specific response to earthquakes and improve assessments of seismic risk at the facilities.[43]

According to the USGS, the 38 proposed stations in seismically active zones would bolster earthquake monitoring where it is currently inadequate. The new stations would fill in where existing stations are farther than 50 kilometers apart (about 31 miles) or where existing stations could go off-scale during an earthquake of expected magnitude for the seismic zone. These stations would also help improve assessments of how well the seismic waves travel, or propagate, from the epicenter outwards, and inversely how the seismic energy is lost, or attenuated, as the earthquake waves move away from the epicenter. As people experienced first-hand, Virginia's August 2011 earthquake broadcast widespread ground shaking. For similar magnitude earthquakes, seismic waves travel more efficiently in general in the CEUS and shaking is felt over a broader area compared to the western United States.

According to its report, the USGS proposes deploying 62 additional stations near nuclear power plants to help characterize earthquake effects for site-specific engineering and emergency-response applications. This would aid in the location and degree of ground shaking and estimation of damage to structures in near-real time.[44] The combined 100 new stations proposed by the USGS would supplement the Advanced National Seismic System (ANSS) to compute earthquake properties more accurately in the region and improve the likelihood of capturing important earthquake data within 50 kilometers of the earthquake epicenter for moderate and large earthquakes.[45] The USGS report indicates that this information would be useful to the NRC.

---

[42] William S. Leith, Harley M. Benz, and Robert B. Hermann, *Improved Earthquake Monitoring in the Central and Eastern United States in Support of Seismic Assessments for Critical Facilities*, Department of the Interior, U.S. Geological Survey, Open-File Report 2011-1101, 2011.

[43] Ibid., p. 11.

[44] An example of these types of near-real time forecasts of the amount of shaking and damage is called ShakeMap. For more information, see CRS Report RL33861, *Earthquakes: Risk, Detection, Warning, and Research*, by Peter Folger.

[45] According to the USGS, "the mission of ANSS is to provide accurate and timely data and information products for seismic events, including their effects on buildings and structures, employing modern monitoring methods and technologies." USGS Earthquake Hazards Program, http://earthquake.usgs.gov/research/monitoring/anss/.

The USGS report also includes cost information for the seismometers, which range from $15,000 to $20,000 to purchase and install "Class A" systems, and about $10,000 for "Class B" systems. Class A systems have higher resolution and other features superior to Class B systems. Operating costs range from $2,000 to $5,000 per year. For 100 new stations, costs to purchase and install Class A systems would range between $1.5 and $2.0 million versus $1.0 million for 100 Class B systems. Operating costs would range between $200,000 to $500,000 per year, depending on location and options for transmitting data from the stations.[46]

---

[46] U.S. Geological Survey, Open-File Report 2011-1101, p. 13.

Appendix A. Magnitude, Intensity, and Seismic Spectrum

Earthquake magnitude is a measure of the strength of the earthquake as determined from seismographic observations. Magnitude is essentially an objective, quantitative measure of an earthquake's size expressed in various ways based on seismographic records (e.g., Richter Local Magnitude, Surface Wave Magnitude, Body Wave Magnitude, and Moment Magnitude).[47] Currently, the most commonly used magnitude measurement is Moment Magnitude (M), which accounts for the strength of the rock that ruptured, the area of the fault that ruptured, and the average amount of slip.[48] Moment is a physical quantity proportional to the slip on the fault times the area of the fault surface that slips. It relates to the total energy released in the earthquake. The moment can be estimated from seismograms (and from geodetic measurements). The Moment Magnitude provides an estimate of earthquake size that is valid over the complete range of magnitudes, a characteristic that was lacking in other magnitude scales, such as the Richter scale.

Because of the logarithmic basis of the moment magnitude scale, each whole number increase in magnitude represents a tenfold increase in measured amplitude; as an estimate of energy, each whole number step in the magnitude scale corresponds to the release of about 31 times more energy than the amount associated with the preceding whole number value.

In 1935, Charles F. Richter of the California Institute of Technology developed the Richter magnitude scale based on the behavior of a specific seismograph manufactured at that time. The instruments are no longer in use and therefore the Richter magnitude scale is no longer used in the technical community. However, the term Richter Scale is so common in use that scientists generally just answer questions about "Richter" magnitude by substituting moment magnitude without correcting the misunderstanding.

The intensity of an earthquake is a qualitative assessment of effects of the earthquake at a particular location. The assigned intensity factors include observed effects on humans, on human-built structures, and on the earth's surface at a particular location. The most commonly used scale in the United States is the Modified Mercalli Intensity (MMI) scale, which has values ranging from I to XII in the order of severity. MMI of I indicates an earthquake that was not felt except by a very few, whereas MMI of XII indicates damage to all works of construction, either partially or completely. While an earthquake has only one magnitude, intensity depends on the effects at each particular location.

Greater magnitude earthquakes are generally associated with greater lengths of fault ruptures.[49] A fault break of 100 miles might be associated with an M8 earthquake, while a break of several miles might generate an M6 earthquake. The length of the fault break, however, is not directly proportional to the energy released. The induced amplitude of acceleration (g) does increase with increasing magnitude (M). Various methods developed relate the magnitude of an earthquake to

---

[47] US NRC, *NRC frequently asked questions related to the March 11, 2011 Japanese Earthquake and Tsunami.*

[48] USGS, *Measuring Earthquakes*, http://earthquake.usgs.gov/learn/faq/?categoryID=2&faqID=23.

[49] H. Bolton Seed, I. M. Idriss, and Fred. W. Kiefer, "Characteristics of Rock Motions During Earthquakes," *Journal of Soil Mechanics and Foundation Division, Proceedings of the American Society of Civil Engineers*, September 1969, pp. 1199-1217.

the amplitude of acceleration it induces, and different methods may result in significant variations in results.

Appendix B. Early Seismic Zone Map

In the late 1940s, structural engineers began considering the seismic-based shear forces that structures must resist. To supplement their design calculations, they referred to the Seismic Zone Map published by the Uniform Building Code (UBC, in **Figure B-1**). The UBC map divided the United States into six distinct seismic zones representing various degrees of seismic risk. The map expressed peak ground acceleration as the decimal ratio of the acceleration due to gravity (g) that applied to a Maximum Credible Earthquake (MCE) and an Operating Basis Earthquake (OBE). UBC defined a maximum credible earthquake as producing the greatest level of ground motion at a site. An operating basis earthquake was be defined as the greatest level of ground motion likely to occur during the economic life of a structure.

Figure B-1. Seismic Zone Map of the United States

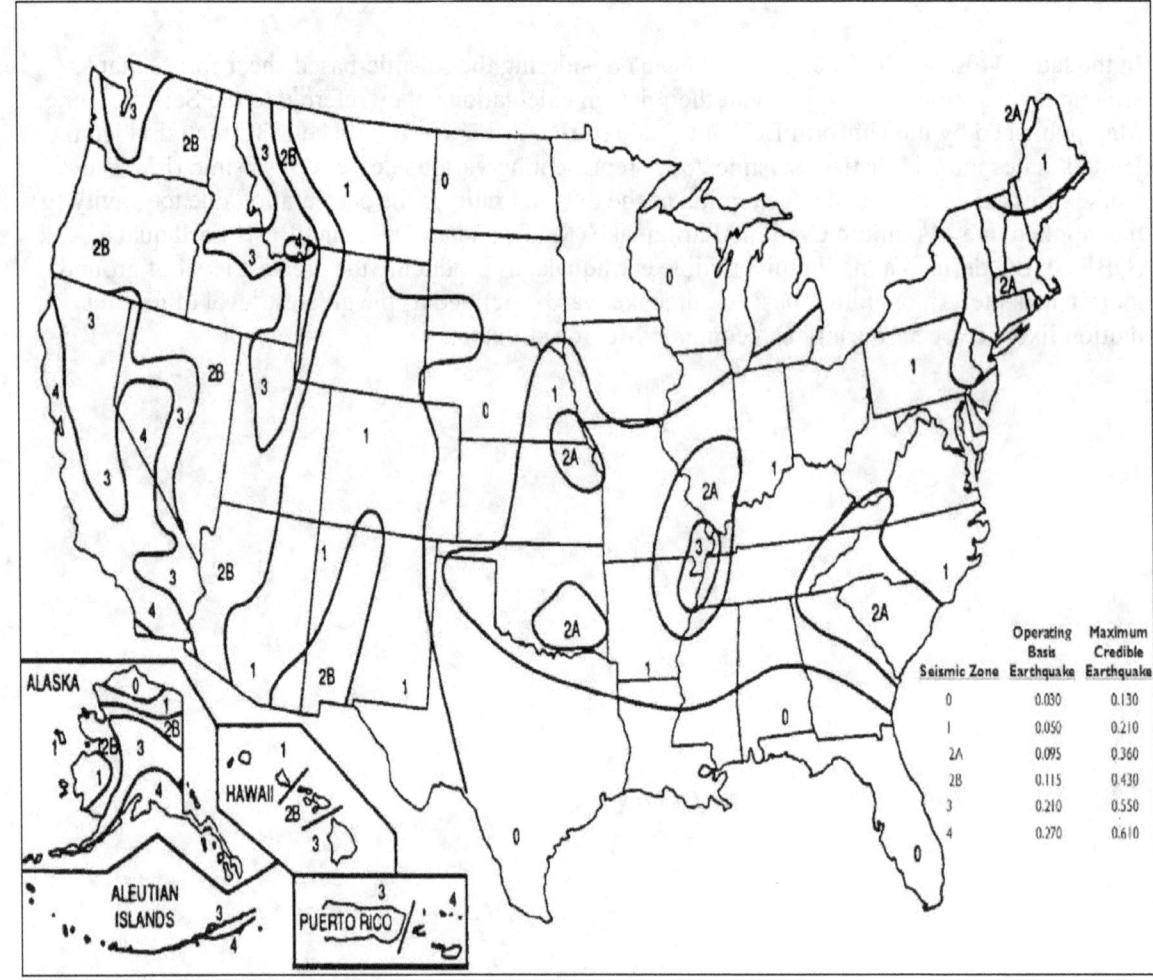

Source: U.S Army Corps of Engineers EP 1110-2-12 (Uniform Building Code, 1988 Edition).

Notes: PGAs are expressed as the decimal ratio of the acceleration of gravity. The PGA for the Operating Basis Earthquake is based on a 50% chance of exceedance in 100 years. The Maximum Credible Earthquake is considered to be an event with a 5,000-year return period (annual risk of exceedance = 0.0002 chance per year).

Appendix C. Terms

Boiling water reactor (BWR) directly generates steam inside the reactor vessel.

Deterministic Seismic Hazard Assessment (DSHA) focuses on a single earthquake event to determine the finite probability of damage occurring.

Double-ended guillotine break (DEGB) represents a break of the largest diameter pipe in the primary system that the emergency core cooling system (ECCS) must be sized to provide adequate makeup water to compensate for.

Light water reactor systems use ordinary water as a fuel moderator and coolant, and uranium fuel artificially enriched to 3%-5% fissile uranium-235. Includes BWR and PWR types.

Loss of Coolant Accident (LOCA) is an accident involving a broken pipe, stuck-open valve, or other leak in the reactor coolant system that results in a loss of the water cooling the reactor core.

Operating Basis Earthquake is the maximum vibratory ground motion for which a reactor could continue operation without undue risk and safety of the public.

Pressurized water reactor (PWR) uses two major loops to convert the heat generated by the reactor core into steam outside of the reactor vessel.

Probabilistic Seismic Hazard Assessments (PSHA) attempt to quantify the probability of exceeding various ground-motion levels at a site given all possible earthquakes.

Safe Shutdown Earthquake (also design basis earthquake) is the maximum vibratory ground motion at which certain structures, systems, and components are designed to remain functional.

Seismic Core Damage Frequency is the probability of damage to the core resulting from a seismic initiating event.

Author Contact Information

Anthony Andrews
Specialist in Energy and Defense Policy
aandrews@crs.loc.gov, 7-6843

Peter Folger
Specialist in Energy and Natural Resources Policy
pfolger@crs.loc.gov, 7-1517

Acknowledgments

Jacqueline C. Nolan, Library of Congress, Geography and Maps Division

Richard J. Campbell, Specialist in Energy Policy, Congressional Research Service

Mark Holt, Specialist in Energy Policy, Congressional Research Service

Jon Ake, NRC Office of Research, Structural, Geotechnical and Seismic Engineering Branch

Cliff Munson, NRC Office of New Reactors , Division of Site and Environmental Reviews